GENERAL RELATIVITY

DEN TAYLOR

Relativity material science

Relativity, far reaching actual hypotheses framed by the German-conceived physicist Albert Einstein. With his hypotheses of unique relativity (1905) and general relativity (1915), Einstein toppled numerous suppositions basic before actual speculations, reclassifying in the process the key ideas of room, time, matter, energy, and gravity. Alongside quantum mechanics, relativity is vital to present day physical science. Specifically, relativity gives the premise to figuring out inestimable cycles and the math of the actual universe.

"Unique relativity" is restricted to objects that are moving as for inertial casings of reference — i.e, in a condition of uniform movement as for each other with the end goal that an onlooker can't, by simply mechanical trials, recognize one from the other. Starting with the way of behaving of light (and any remaining electromagnetic radiation), the hypothesis of exceptional relativity makes inferences that are in opposition to regular experience however completely affirmed by tests. Unique relativity uncovered that the speed of light is a breaking point that can be drawn nearer yet not arrived at by any material item; it is the beginning of the most popular condition in science, E = mc2; and it has prompted other tempting results, for example, the "twin oddity."

"General relativity" is worried about gravity, one of the essential powers known to mankind. (The others are electromagnetism, the solid power, and the frail power.) Gravity characterizes naturally visible way of behaving, thus broad relativity portrays huge scope actual peculiarities like planetary elements, the birth and passing of stars, dark openings, and the development of the universe.

Exceptional and general relativity have significantly impacted actual science and human life, most decisively in uses of thermal power and atomic weapons. Moreover, relativity and its reexamining of the principal classes of existence have given a premise to specific philosophical, social, and imaginative translations that have impacted human culture in various ways.

Cosmology before relativity

Relativity changed the logical origination of the universe, which started in endeavors to get a handle on the powerful way of behaving of issue. In Renaissance times, the incomparable Italian physicist Galileo Galilei moved past Aristotle's way of thinking to present the advanced investigation of mechanics, which requires quantitative estimations of bodies moving in existence. His work and that of others prompted fundamental ideas, for example, speed, which is the distance a body covers in a provided guidance for every unit time; speed increase, the pace of progress of speed; mass, how much material in a body; and power, a push or pull on a body.

The following significant step happened in the late seventeenth hundred years, when the English logical virtuoso Isaac Newton formed his three popular laws of movement, the first and second are of extraordinary worry in relativity. Newton's most memorable regulation, known as the law of dormancy, expresses that a body that isn't followed up on by outer powers goes through no speed increase — either staying very still or proceeding to move in an orderly fashion at steady speed. Newton's subsequent regulation expresses that a power applied to a body changes speed by creating a speed increase is relative to the power and contrarily corresponding to the mass of the body. In developing his framework, Newton additionally characterized reality, taking both to be absolutes that are unaffected by anything outside. Time, he expressed, "streams equably," while space "remains generally comparable and relentless."

Newton's regulations demonstrated legitimate in each application, as in computing the way of behaving of falling bodies, yet they likewise gave the system to his milestone law of gravity (the term, got from the Latin gravis, or "weighty," had been being used since basically the sixteenth 100 years). Starting with the (maybe legendary) perception of a falling apple and afterward taking into account the Moon as it circles Earth, Newton reasoned that an undetectable power acts between the Sun and its planets. He figured out a similarly basic numerical articulation for the gravitational power; it expresses that each article in the universe draws in each and every item with a power that works through void space and that shifts with the majority of the items and the distance between them.

The law of gravity was splendidly fruitful in making sense of the system behind Kepler's laws of planetary movement, which the German stargazer Johannes Keller had figured out toward the start of the seventeenth hundred years. Newton's mechanics and law of gravity, alongside his presumptions about the idea of reality, appeared to find success in making sense of the elements of the universe, from movement on Earth to astronomical occasions.

Light and the ether

In any case, this accomplishment at making sense of regular peculiarities came to be tried from an unforeseen heading — the way of behaving of light, whose elusive nature had perplexed rationalists and researchers for quite a long time. In 1865 the Scottish physicist James Representative Maxwell showed that light is an electromagnetic wave with swaying electrical and attractive parts. Maxwell's conditions anticipated that electromagnetic waves would go through void space at a speed of precisely 3 × 108 meters each second (186,000 miles each second) — i.e., concurring with the deliberate speed of light. Explores before long affirmed the electromagnetic idea of light and laid out its speed as a key boundary of the universe.

Maxwell's striking outcome responded to well established inquiries regarding light, however it raised another central issue: on the off chance that light is a moving wave, what medium backings it? Sea waves and sound waves comprise of the ever-evolving oscillatory movement of particles of water and of air gases, separately. However, would could it be that vibrates to make a moving light wave? Or on the other hand to put it another way, how does the energy typified in light travel from one highlight another?

For Maxwell and different researchers of the time, the response was that light gone in a speculative medium called the ether (aether). Evidently, this medium pervaded all space without blocking the movement of planets and stars; yet it must be more unbending than steel so that light waves could travel through it at rapid, similarly that a tight guitar string upholds quick mechanical vibrations. In spite of this inconsistency, the possibility of the ether appeared to be fundamental — until a conclusive examination discredited it.

In 1887 the German-conceived American physicist A.A. Michelson and the American scientist Edward Morley made flawlessly exact estimations to decide what Earth's movement through the ether meant for the deliberate speed of light. In old style mechanics, Earth's development would add to or deduct from the deliberate speed of light waves, similarly as the speed of a boat would add to or take away from the speed of sea waves as estimated from the boat. In any case, the Michelson-Morley explore had a surprising result, for the deliberate speed of light continued as before no matter what Earth's movement. This must be that the ether had no significance and that the way of behaving of light couldn't be made sense of by traditional material science. The clarification arose, all things considered, from Einstein's hypothesis of unique relativity.

Exceptional relativity

Researchers, for example, Austrian physicist Ernst Mach and French mathematician Henri Poincare had scrutinized traditional mechanics or examined the way of behaving of light and the importance of the ether before Einstein. Their endeavors gave a foundation to Einstein's novel way to deal with understanding the universe, which he brought in his local German a Gedankenexperiments, or "psychological test."

Einstein portrayed how at age 16 he watched himself to his eye as he rode on a light wave and looked at another light wave moving lined up with his. As per old style material science, Einstein ought to have seen the subsequent light wave moving at an overall speed of nothing. Notwithstanding, Einstein realize that Maxwell's electromagnetic conditions totally expect that light generally move at 3×108 meters each second in a vacuum. Nothing in the hypothesis permits a light wave to have a speed of nothing. One more issue emerged too: assuming that a decent spectator views light as having a speed of 3×108 meters each second, while an eyewitness moving at the speed of light views light as having a speed of nothing, it would imply that the laws of electromagnetism rely upon the onlooker. Yet, in old style mechanics similar regulations apply for all spectators, and Einstein saw no great explanation for why the electromagnetic regulations ought not be similarly general. The steadiness of the speed of light and the all-inclusiveness of the laws of physical science for all eyewitnesses are foundations of unique relativity.

Beginning stages and proposes

In creating unique relativity, Einstein started by tolerating what examination and his own reasoning demonstrated to be the genuine way of behaving of light, in any event, when this went against traditional material science or the standard discernments about the world.

The way that the speed of light is no different for all spectators is peculiar in common terms. In the event that a traveler in a train moving at 100 km each hour shoots a bolt in the train's bearing of movement at 200 km each hour, a trackside spectator would gauge the speed of the bolt as the amount of the two velocities, or 300 km each hour. In similarity, assuming the train moves at the speed of light and a traveler sparkles a laser in a similar bearing, then, at that point, sound judgment shows that a trackside spectator ought to see the light moving at the amount of the two velocities, or two times the speed of light (6 × 108 meters each second).

While such a law of expansion of speeds is legitimate in traditional mechanics, the Michelson-Morley try showed that light doesn't comply with this regulation. This goes against sound judgment; it suggests, for example, that both a train moving at the speed of light and a light bar radiated from the train show up at a point.

Outcomes of the hypothesizes

Relativistic reality

To make the speed of light steady, Einstein supplanted outright existence with new definitions that rely upon the condition of movement of an onlooker. Einstein made sense of his methodology by thinking about two eyewitnesses and a train. One spectator remains close by a straight track; different rides a train moving at consistent speed along the track. Each perspectives the world comparative with his own environmental elements. The decent eyewitness estimates distance from an imprint recorded on the track and measures time with his watch; the train traveler estimates distance from an imprint engraved on his railroad vehicle and measures time with his own watch.

On the off chance that time streams something very similar for the two eyewitnesses, as Newton accepted, the two casings of reference are accommodated by the connection: $x' = x - vt$. Here x is the distance to some particular occasion that occurs along the track, as estimated by the decent eyewitness; x' is the distance to a similar occasion as estimated by the moving onlooker; v is the speed of the train — that is, the speed of one spectator comparative with the other; and t is the time at which the occasion occurs, something similar for the two onlookers. For instance, assume the train moves at 40 km each hour. One hour after it sets out, a tree 60 km from the train's beginning stage is struck by lightning. The proper eyewitness estimates x as 60 km and t as 60 minutes. The moving spectator additionally gauges t as 60 minutes, thus, as per Newton's condition; he gauges x' as 20 km.

This examination appears glaringly evident; however Einstein saw a nuance concealed in its hidden presumptions — specifically, the issue of synchronization. The two individuals don't really notice the lightning strike simultaneously. Indeed, even at the speed of light, the picture of the strike gets some margin to arrive at every onlooker, and, since each is at an alternate separation from the occasion, the movement times contrast. Taking this understanding further, assume lightning strikes two trees, one 60 km in front of the decent onlooker and the other 60 km behind, precisely as the moving spectator passes the proper eyewitness. Each picture ventures to every part of a similar distance to the proper onlooker, thus he positively witnesses the occasions all the while. The movement of the drawing eyewitness brings him nearer to one occasion than the other, in any case, and he subsequently witnesses the occasions at various times.

Einstein inferred that synchronization is relative; occasions that are synchronous for one eyewitness may not be for another. This drove him to the unreasonable thought that time streams diversely as per the condition of movement and to the end that distance is likewise relative. In the model, the train traveler and the decent eyewitness can each stretch a measuring tape from back to front of a railroad vehicle to track down its length. The two closures of the tape should be set ready at a similar moment — that is, at the same time — to get a genuine worth. Notwithstanding, in light of the fact that the importance of concurrent is different for the two eyewitnesses, they measure various lengths.

This thinking drove Einstein to new conditions for reality, called the Lorentz changes, after the Dutch physicist Hendrik Lorentz, who previously proposed them. They are:

Where t' is time as estimated by the moving spectator and c is the speed of light.

From these situations, Einstein determined another relationship that replaces the traditional law of expansion of speeds,

Einstein's speed expansion

Where u and u′ are the speed of any moving article as seen by every onlooker and v is again the speed of one spectator comparative with the other. This connection ensures Einstein's most memorable hypothesize (that the speed of light is steady for all onlookers). On account of the spotlight pillar projected from a train moving at the speed of light, an eyewitness on the train estimates the speed of the shaft as c. As per the condition above, so does the trackside eyewitness, rather than the worth 2c that traditional material science predicts.

To make the speed of light steady, the hypothesis expects that reality change in a moving body, as per its speed, as seen by an external onlooker. The body becomes more limited along its bearing of movement; that is, its length contracts. Time spans become longer, implying that time runs all the more leisurely in a moving body; that is, time enlarges. In the train model, the individual close to the track estimates a more limited length for the train and a more extended time stretch for tickers on the train than does the train traveler. The relations portraying these progressions are

Relativistic length-time

Where L0 and T0, called legitimate length and appropriate time, individually, are the qualities estimated by a spectator on the moving body, and L and T are the relating amounts as estimated by a proper eyewitness.

The relativistic impacts become enormous at speeds close to that of light, despite the fact that it is important again that they show up just when an onlooker checks a moving body out. He never sees changes in space or time inside his own reference outline (whether on a train or space apparatus), even at the speed of light. These impacts don't show up in normal life, in light of the fact that the component $v2/c2$ is minute at even the most noteworthy rates achieved by people, so that Einstein's conditions become basically equivalent to the traditional ones.

To determine further outcomes, Einstein joined his redefinitions of existence with two strong actual standards: protection of energy and preservation of mass, which express that the aggregate sum of each stays steady in a shut framework. Einstein's subsequent propose guaranteed that these regulations stayed legitimate for all eyewitnesses in the new hypothesis, and he utilized them to determine the relativistic implications of mass and energy.

One outcome is that the mass of a body speeds up. An eyewitness on a moving body, for example, a space apparatus, gauges its purported rest mass $m0$, while a proper spectator estimates its mass m as

Relativistic mass

Which is more prominent than $m0$? Truth be told, as the shuttle's speed moves toward that of light, the mass m methodologies boundlessness. Nonetheless, as the item's mass increments, so does the energy expected to continue to speed up it; consequently, it would take endless energy to speed up a material body to the speed of light. Thus, no material article can arrive at the speed of light, which is as far as possible for the universe. (Light itself can achieve this speed on the grounds that the rest mass of a photon, the quantum molecule of light, is zero.)

$E = mc^2$

Einstein's treatment of mass showed that the expanded relativistic mass comes from the energy of movement of the body — that is, its dynamic energy E — separated by c2. This is the beginning of the renowned condition $E = mc^2$, which communicates the way that mass and energy are a similar actual substance and can be changed into one another.

The twin mystery

The unreasonable idea of Einstein's thoughts makes them hard to retain and brings about circumstances that appear to be inconceivable. One notable case is the twin Catch 22, an appearing irregularity in how extraordinary relativity portrays time.

Assume that one of two indistinguishable twin sisters takes off into space at almost the speed of light. As per relativity, time runs more leisurely on her rocket than on The planet; consequently, when she gets back to Earth, she will be more youthful than her Earth-bound sister. In any case, in relativity, what one eyewitness sees as happening to a subsequent one, the subsequent one sees as happening to the first. To the space-going sister, time moves more leisurely on Earth than in her shuttle; when she returns, her Earth-bound sister is the person who is more youthful. How could the space-going twin be both more youthful and more seasoned than her Earth-bound sister?

The response is that the mystery is just clear, for the circumstance isn't properly treated by exceptional relativity. To get back to Earth, the space apparatus should take an alternate route, which abuses the state of consistent straight-line movement vital to exceptional relativity. A full treatment requires general relativity, which shows that there would be an uneven change in time between the two sisters. In this manner, the "mystery" doesn't raise serious questions about how exceptional relativity portrays time, which has been affirmed by various examinations.

Exceptional relativity is less unmistakable than old style physical science in that both the distance D and time span T between two occasions rely upon the spectator. Einstein noted, nonetheless, that a specific blend of D and T, the amount D2 − c2T2, has similar incentive for all onlookers.

The term CT in this invariant amount raises time to a sort of numerical equality with space. Taking note of this, the German numerical physicist Hermann Murkowski showed that the universe looks like a four-layered structure with facilitates x, y, z, and ct addressing length, width, level, and time, separately. Subsequently, the universe can be depicted as a four-layered space-time continuum, a focal idea in everyday relativity.

Exploratory proof for exceptional relativity

Since relativistic changes are little at normal paces for perceptible articles, the affirmation of unique relativity has depended on either the assessment of subatomic bodies at high rates or the estimation of little changes by touchy instrumentation. For instance, super exact clocks were put on various business aircrafts flying at one-millionth the speed of light. Following two days of constant flight, the time shown by the airborne clocks contrasted by parts of a microsecond from that shown by a synchronized clock left on The planet, as anticipated.

Bigger impacts are seen with rudimentary particles moving at speeds near that of light. One such test included muons, rudimentary particles made by grandiose beams in Earth's air at an elevation of around 9 km (30,000 feet). At 99.8 percent of the speed of light, the muons ought to arrive adrift level in 31 microseconds, yet estimations showed that it took just 2 microseconds. That's what the explanation is, comparative with the moving muons, the distance of 9 km contracted to 0.58 km (1,900 feet). Likewise, a relativistic mass increment has been affirmed in estimations on quick rudimentary particles, where the change is huge (see beneath Molecule gas pedals).

Such outcomes leave most likely that unique relativity accurately depicts the universe, albeit the hypothesis is hard to acknowledge at an instinctive level. Some knowledge comes from Einstein's remark that in relativity the restricting rate of light assumes the part of a limitless speed. At boundless speed, light would navigate any distance in zero time. Essentially, as per the relativistic conditions, an eyewitness riding a light wave would see lengths agreement to nothing and timekeepers quit ticking as the universe moved toward him at the speed of light. Actually, relativity replaces a boundless speed limit with the limited worth of 3 × 108 meters each second.

General relativity

Foundations of general relativity

Since Isaac Newton's law of gravity served so well in making sense of the way of behaving of the planetary group, the inquiry emerges why fostering another hypothesis of gravity was important. The response is that Newton's hypothesis disregards extraordinary relativity, for it requires a vague "activity a good ways off" through which any two articles — like the Sun and Earth — quickly pull one another, regardless of how far separated. Notwithstanding, immediate reaction would require the gravitational communication to spread at boundless speed, which is blocked by exceptional relativity. By and by, this is no extraordinary issue for depicting our nearby planet group, for Newton's regulation offers substantial responses for objects moving gradually contrasted and light. By and by, since Newton's hypothesis can't be thoughtfully accommodated with unique relativity, Einstein went to the improvement of general relativity as a better approach to figure out attraction.

Guideline of identicalness

To start fabricating his hypothesis, Einstein held onto on an understanding that came to him in 1907. As he made sense of in a talk in 1922:

I was perched on a seat in my patent office in Bern. Out of nowhere an idea struck me: On the off chance that a man falls uninhibitedly, he wouldn't feel his weight. I was shocked. This basic psychological test established a profound connection with me. This drove me to the hypothesis of gravity.

Einstein was suggesting an inquisitive reality known in Newton's time: regardless of what the mass of an item, it falls toward Earth with a similar speed increase (overlooking air opposition) of 9.8 meters each second squared. Newton made sense of this by hypothesizing two sorts of mass: inertial mass, which opposes movement and goes into his overall laws of movement, and gravitational mass, which goes into his situation for the power of gravity. That's what he showed, in the event that the two masses were equivalent; all items would fall with that equivalent gravitational speed increase.

Einstein, in any case, acknowledged something more significant. An individual remaining in a lift with a wrecked link feels weightless as the nook falls uninhibitedly toward Earth. The explanation is that both he and the lift advance quickly descending at a similar rate thus fall at the very same speed; consequently, shy of looking external the lift at his environmental factors, he can't confirm that he is being pulled descending. As a matter of fact, there is no examination he can do inside a fixed falling lift to confirm that he is inside a gravitational field. In the event that he sets a ball free from his hand, it will fall at a similar rate, basically remaining where he delivers it. What's more, if he somehow happened to see the ball sink toward the floor, he was unable to let know if that was on the grounds that he was very still inside a gravitational field that pulled the ball down or in light of the fact that a link was yanking the lift up so that its floor rose toward the ball.

Einstein communicated these thoughts in his misleading straightforward standard of proportionality, which is the premise of general relativity: on a nearby scale — significance inside a given framework, without taking a gander at different frameworks — it is difficult to recognize actual impacts because of gravity and those because of speed increase.

All things considered, proceeded with Einstein's Gedankenexperiments, light should be impacted by gravity. Envision that the lift has an opening worn straight through two inverse walls out. At the point when the lift is very still, a light emission entering one opening goes in an orderly fashion lined up with the floor and exits through the other opening. Be that as it may, in the event that the lift is advanced quickly vertically, when the beam arrives at the subsequent opening, the opening has moved and is not generally lined up with the beam. As the traveler sees the light miss the subsequent opening, he reasons that the beam has followed a bended way (as a matter of fact, a parabola).

On the off chance that a light beam is twisted in a sped up framework, as per the standard of proportionality, light ought to likewise be bowed by gravity, going against the ordinary assumption that light will go in an orderly fashion (except if it passes starting with one medium then onto the next). In the event that its way is bended by gravity that should really intend that "straight line" has an alternate significance close to a gigantic gravitational body, for example, a star than it truly does in purge space. This was a clue that gravity ought to be treated as a mathematical peculiarity.

Bended space-time and mathematical attraction

The particular component of Einstein's perspective on gravity is its mathematical nature. (See likewise calculation: this present reality.) While Newton felt that gravity was a power, Einstein showed that gravity emerges from the state of room time. While this is challenging to envision, there is a relationship that gives some knowledge — despite the fact that it is an aide, as opposed to a conclusive assertion of the hypothesis.

The relationship starts by considering space-time as an elastic sheet that can be disfigured. In any district far off from gigantic vast items, for example, stars, space-time is uncurved — that is, the elastic sheet is totally level. If one somehow happened to test space-time around there by conveying a beam of light or a test body, both the beam and the body would go in completely straight lines, similar to a kid's marble moving across the elastic sheet.

Be that as it may, the presence of a gigantic body bends space-time, as though a bowling ball were put on the elastic sheet to make a cuplike misery. In the similarity, a marble set close to the downturn rolls down the slant toward the bowling ball as though pulled by a power. What's more, in the event that the marble is given a sideways push, it will depict a circle around the bowling ball, as though a consistent draw toward the ball is swinging the marble into a shut way.

Along these lines, the ebb and flow of room time close to a star characterizes the most brief regular ways, or geodesics — much as the most brief way between any two focuses on Earth is certainly not a straight line, which can't be built on that bended surface, however the circular segment of an extraordinary circle course. In Einstein's hypothesis, space-time geodesics characterize the diversion of light and the circles of planets. As the American hypothetical physicist John Wheeler put it, matter tells space-time how to bend, and space-time advises matter how to move.

The arithmetic of general relativity

The elastic sheet similarity assists with representation of room time, however Einstein himself fostered a total quantitative hypothesis that portrays space-time through exceptionally dynamic science. General relativity is communicated in a bunch of interlinked differential conditions that characterize how the state of room time relies upon how much matter (or, equally, energy) in the locale. The arrangement of these supposed field conditions can yield replies to various actual circumstances, including the way of behaving of individual bodies and of the whole universe. Einstein quickly comprehended that the field conditions could depict the whole universe. In 1917 he changed the first variant of his situations by adding what he called the "cosmological term." This addressed a power that acted to make the universe extend, subsequently neutralizing gravity, which will in general make the universe contract. The outcome was a static universe, as per the best information on the time.

In 1922, nonetheless, the Soviet mathematician Aleksandra Aleksandrovich Friedman showed that the field conditions foresee a unique universe, which can either grow perpetually or go through patterns of exchanging extension and constriction. Einstein came to concur with this outcome and deserted his cosmological term. Later work, prominently spearheading estimations by the American space expert Edwin Hubble and the advancement of the enormous detonation model, has affirmed and intensified the idea of a growing universe.

In 1916 the German cosmologist Karl Schwarzschild utilized the field conditions to work out the gravitational impact of a solitary circular body like a star. On the off chance that the mass is neither extremely huge nor profoundly thought, the subsequent estimation will be equivalent to that given by Newton's hypothesis of gravity. Consequently, Newton's hypothesis isn't erroneous; rather, it is a legitimate guess to general relativity under specific circumstances.

Schwarzschild likewise depicted another impact. In the event that the mass is moved in a vanishingly little volume — a peculiarity — gravity will turn out to areas of strength for be the point that nothing maneuvered into the encompassing district can at any point leave. Indeed, even light can't get away. In the elastic sheet similarity, maybe a little huge article makes a downturn so steep that nothing can get away from it. In acknowledgment that this extreme space-time contortion would be imperceptible — on the grounds that it would retain light and never discharge any — it was named a dark opening.

In quantitative terms, Schwarzschild's outcome characterizes a circle that is focused at the peculiarity and whose sweep relies upon the thickness of the encased mass. Occasions inside the circle are perpetually detached from the rest of the universe; hence, the Schwarzschild sweep is known as the occasion skyline. No human innovation could minimized matter adequately to make dark openings, however they happen as conclusive strides in the existence pattern of stars. All following millions or billions of years, a star goes through hydrogen and different components produce energy through atomic combination. With its atomic heater banked, the star no longer keeps an interior strain to extend, and gravity is left unopposed to pull internal and pack the star. For stars over a specific mass, this gravitational breakdown will deliver a dark opening containing a few times the mass of the Sun. In different cases, the gravitational breakdown of gigantic residue mists can make supermassive dark openings containing millions or billions of sun based masses.

Astrophysicists have found numerous enormous items that contain such a thick centralization of mass in a little volume. These dark openings incorporate one at the focal point of the Smooth Way System (Sagittarius A*) and certain twofold stars that emanate X-beams as they circle one another. One, at the focal point of the system M87, has even been straightforwardly imaged. The hypothesis of dark openings has prompted one more anticipated substance, a wormhole. This is an answer of the field conditions that looks like a passage between two dark openings or different focuses in space-time. Such a passage would give an easy route between its end focuses. In similarity, consider a subterranean insect strolling across a level piece of paper from point A to point B. In the event that the paper is bended through the third aspect, so that An and B cross-over, the subterranean insect can step straightforwardly from one highlight the other, hence staying away from a long trip.The chance of short-circuiting the huge distances between stars makes wormholes alluring for space travel. Since the passage joins minutes in time as well as areas in space, it likewise has been contended that a wormhole would permit travel into the past. Notwithstanding, wormholes are inherently unsound. While intriguing adjustment plans have been proposed, there is at this point no proof that these can work or for sure that wormholes exist.

Test proof for general relativity

Not long after the hypothesis of general relativity was distributed in 1915, the English cosmologist Arthur Edenton considered Einstein's expectation that light beams are bowed close to a monstrous body, and he understood that it very well may be checked via cautiously looking at star positions in pictures of the Sun taken during a sun oriented obscure with pictures of a similar district of room taken when the Sun was in an alternate part of the sky. Confirmation was postponed by The Second Great War, yet in 1919 an amazing open door introduced itself with a particularly lengthy absolute sun oriented obscure, nearby the splendid Hyades star group, that was noticeable from northern Brazil to the African coast. Eddington drove one undertaking to Príncipe, an island off the African coast, and Andrew Crommelin of the Regal Greenwich Observatory drove a second endeavor to Sobral, Brazil. After cautiously contrasting photos from the two campaigns and reference photos of the Hyades, Eddington pronounced that the starlight had been avoided around 1.75 seconds of circular segment, as anticipated by broad relativity. (A similar impact produces gravitational lensing, where a huge grandiose article shines light from one more article past it to create a contorted or amplified picture. The cosmic revelation of gravitational focal points in 1979 gave extra help for general relativity.)

Additional proof came from the planet Mercury. In the nineteenth 100 years, it was found that Mercury doesn't get back to the very same spot each time it finishes its curved circle. All things being equal, the circle pivots gradually in space, so that on each circle the perihelion — the place of nearest way to deal with the Sun — moves to a marginally unique point. Newton's law of gravity couldn't make sense of this perihelion shift, yet broad relativity gave the right circle.

One more affirmed expectation of general relativity is that time enlarges in a gravitational field, implying that tickers run more slow as they approach the mass that is delivering the field. This has been estimated straightforwardly and furthermore through the gravitational redshift of light. Time expansion makes light vibrate at a lower recurrence inside a gravitational field; subsequently, the light is moved toward a more drawn out frequency — that is, close to the red. Different estimations have checked the comparability rule by showing that inertial and gravitational mass is unequivocally something similar. The most striking expectation of general relativity is that of gravitational waves. Electromagnetic waves are brought about by sped up electrical charges and are distinguished when they set up different charges. Additionally, gravitational waves would be brought about by masses moving and are recognized when they start movement in different masses. Notwithstanding, gravity is exceptionally feeble contrasted and electromagnetism. Just an immense vast occasion, like the crash of two stars, can create distinguishable gravitational waves. Endeavors to detect gravitational waves started during the 1960s, and such waves were first distinguished in 2015 when LIGO noticed two dark openings 1.3 million light-years away spiraling into one another.

Uses of relativistic thoughts

Albeit relativistic impacts are immaterial in normal life, relativistic thoughts show up in a scope of regions from crucial science to regular citizen and military innovation.

Rudimentary particles

The relationship $E = mc^2$ is fundamental in the investigation of subatomic particles. It decides the energy expected to make particles or to change over one sort into another and the energy delivered when a molecule is destroyed. For instance, two photons, every one of energy E, can crash to frame two particles, each with mass m = E/c2. This pair-creation process is one stage in the early advancement of the universe, as portrayed in the enormous detonation model.

Molecule gas pedals

Information on rudimentary particles comes essentially from molecule gas pedals. These machines raise subatomic particles, generally electrons or protons, to almost the speed of light. At the point when these vivacious shots crush into chosen targets, they explain how subatomic particles associate and frequently produce new types of rudimentary particles.

Molecule gas pedals couldn't be as expected planned without extraordinary relativity. In the sort called an electron synchrotron, for example, electrons gain energy as they navigate a colossal round raceway. At scarcely underneath the speed of light, their mass is huge number of times bigger than their rest mass. Subsequently, the attractive field used to hold the electrons in round circles should be large number of times more grounded than if the mass didn't change.

Parting and combination: bombs and heavenly cycles

Energy is delivered in two sorts of atomic cycles. In atomic parting a weighty core, like uranium, parts into two lighter cores; in atomic combination two light cores consolidate into a heavier one. In each cycle the complete last mass is not exactly the beginning mass. The distinction shows up as energy as per the connection $E = \Delta mc^2$, where Δm is the mass deficiency.

Splitting is utilized in nuclear bombs and in reactors that produce power for regular citizen and military applications. The combination of hydrogen into helium is the energy source in stars and gives the force of a nuclear bomb. Endeavors are currently under method for creating controllable hydrogen combination as a spotless, plentiful power source.

The worldwide situating framework

The worldwide situating framework (GPS) relies upon relativistic standards. A GPS recipient decides its area on Earth's surface by handling radio transmissions from at least four satellites. The distance to each satellite is determined as the result of the speed of light and the delay among transmission and gathering of the sign. Nonetheless, Earth's gravitational field and the movement of the satellites cause time-expansion impacts, and Earth's revolution likewise has relativistic ramifications. Thus, GPS innovation incorporates relativistic amendments that empower positions to be determined to inside a few centimeters.

Sign and beginning of the universe, is personally associated with gravity, which decides the perceptible way of behaving of all matter. General relativity plays had an impact in cosmology since the early estimations of Einstein and Friedman. From that point forward, the hypothesis has given a structure to obliging observational outcomes, for example, Hubble's disclosure of the extending universe in 1929, as well as the huge explosion model, which is the for the most part acknowledged clarification of the beginning of the universe.

The most recent arrangements of Einstein's field conditions rely upon explicit boundaries that describe the destiny and state of the universe. One is Hubble's consistent, which characterizes how quickly the universe is extending; the other is the thickness of issue known to man, which decides the strength of gravity. Under a specific basic thickness, gravity would be frail enough that the universe would extend perpetually, so that space would be limitless. Over that worth, gravity would be sufficiently able to make the universe contract back to its unique moment size after a limited time of extension, a cycle called the "huge mash." For this situation, space would be restricted or limited like the outer layer of a circle. Current endeavors in observational cosmology center around estimating the most reliable potential upsides of Hubble's consistent and of basic thickness.

Relativity, quantum hypothesis, and bound together speculations

Vast conduct on the greatest scale is depicted by broad relativity. Conduct on the subatomic scale is depicted by quantum mechanics, which started with crafted by the German physicist Max Planck in 1900 and treats.

After Einstein created relativity, he fruitlessly looked for a supposed bound together field hypothesis with a space-time math that would incorporate every one of the central powers. Different scholars have endeavored to consolidate general relativity with quantum hypothesis, yet the two methodologies treat powers in essentially various ways. In quantum hypothesis, powers emerge from the trade of specific rudimentary particles, not from the state of room time. Moreover, quantum impacts are remembered to cause a serious mutilation of room time at an incredibly limited scale called the Planck length, which is a lot more modest than the size of rudimentary particles. This recommends that quantum gravity can't be perceived without treating space-time at unfathomable scales. Albeit the association between broad relativity and quantum mechanics stays slippery, some headway has been made toward a completely brought together hypothesis. During the 1960s, the electroweak hypothesis gave incomplete unification, showing a typical reason for electromagnetism and the powerless power inside quantum hypothesis. Late examination recommends that superstring hypothesis, where rudimentary particles are addressed not as numerical focuses but rather as minuscule strings vibrating in at least 10 aspects, shows guarantee for supporting total unification, including attractive energy. In any case, until affirmed by trial results, superstring hypothesis will stay an untested speculation.

Scholarly and social effect of relativity

Responses in everyday culture

The effect of relativity has not been restricted to science. Extraordinary relativity showed up on the scene toward the start of the twentieth hundred years, and general relativity turned out to be commonly known after The Second Great War — times when another reasonableness of "innovation" was becoming characterized in workmanship and writing. Furthermore, the affirmation of general relativity given by the sun powered overshadowing of 1919 got wide exposure. Einstein's 1921 Nobel Prize for Physical science (granted for his work on the photon idea of light), as well as the famous insight that relativity was complicated to such an extent that couple of could get a handle on it, immediately transformed Einstein and his hypotheses into social symbols.

The thoughts of relativity were broadly applied — and twisted — not long after their coming. A few scholars deciphered the hypothesis as significance just that everything is relative, and they utilized this idea in fields far off from material science. The Spanish humanist savant and writer José Ortega y Gusset, for example, wrote in The Advanced Topic (1923),

The hypothesis of Einstein is a brilliant evidence of the agreeable variety of all potential perspectives. On the off chance that the thought is reached out to ethics and feel, we will come to encounter history and life in another manner. The progressive part of Einstein's thinking was additionally taken advantage of, as by the American workmanship pundit Thomas Cowardly, who in 1921 looked at the break among old style and current craftsmanship to the break among Newtonian and Einsteinian thoughts regarding reality.

A few saw explicit relations among relativity and craftsmanship emerging from the possibility of a four-layered space-time continuum. In the nineteenth hundred years, improvements in calculation prompted well known interest in a fourth spatial aspect, envisioned as some way or another lying at right points to every one of the three of the normal elements of length, width, and level. Edwin Abbott's Flatland (1884) was the principal famous show of these thoughts. Different works of imagination that followed discussed the final aspect as a field separated from common presence.

Einstein's four-layered universe, with three spatial aspects and one of time, is reasonably not quite the same as four spatial aspects. Yet, the two sorts of four-layered world became conflated in deciphering the new craft of the twentieth hundred years. Early Cubist works by Pablo Picasso that all the while depicted all sides of their subjects became associated with the possibility of higher aspects in space, which a few journalists endeavored to connect with relativity. In 1949, for instance, the craftsmanship student of history Paul LaPorte composed that "the new pictorial expression made by [C]ubism is generally sufficiently made sense of by applying to it the idea of the space-time continuum." Einstein explicitly dismissed this view, saying, "This new creative 'language' shares nothing practically speaking with the Hypothesis of Relativity." By and by, a few specialists unequivocally investigated Einstein's thoughts. In the new Soviet Association of the 1920s, for instance, the writer and artist Vladimir Mayakovsky, a pioneer behind the creative development called Russian Futurism, or Supremacist, employed a specialist to make sense of relativity for him.

The boundless general interest in relativity was reflected in the quantity of books written to explain the subject for nonexperts. Einstein's famous work of exceptional and general relativity showed up very quickly, in 1916, and his article on space-time showed up in the thirteenth release of Encyclopedia Britannica in 1926. Different researchers, like the Russian mathematician Aleksandra Friedmann and the

English stargazer Arthur Eddington, composed famous books regarding the matters during the 1920s. Such books kept on seeming many years after the fact.

At the point when relativity was first declared, general society was commonly awestruck by its intricacy, a legitimate reaction to the mind boggling science of general relativity. Yet, the theoretical, no visceral nature of the hypothesis likewise created responses against its obvious infringement of good judgment. These responses incorporated a political suggestion; in certain quarters, it was thought of as undemocratic to present or support a hypothesis that couldn't be promptly grasped by the normal individual.

In contemporary use, general culture has acknowledged the thoughts of relativity — the difficulty of quicker than-light travel, $E = mc2$, time widening and the twin mystery, the growing universe, and dark openings and wormholes — to the place where they are promptly perceived in the media and give plot gadgets to works of sci-fi. A portion of these thoughts have acquired importance past their stringently logical ones; in the business world, for example, "dark opening" can mean an unrecoverable monetary channel.

Philosophical contemplations

In 1925 the English thinker Bertrand Russell, in his ABC of Relativity, proposed that Einstein's work would prompt new philosophical ideas. Relativity affects reasoning, enlightening an issues that return to the old Greeks. The possibility of the ether, summoned in the late nineteenth 100 years to convey light waves, looks back to Aristotle. He isolated the world into earth, air, fire, and water, with the ether (aether) as the fifth component addressing the unadulterated divine circle. The Michelson-Morley examination and relativity wiped out the last remnants of this thought.

Relativity additionally changed the significance of math as it was created in Euclid's Components (c. 300 BCE). Euclid's framework depended on the maxim "a straight line is the briefest distance between two focuses," among others that appeared to be self-obviously evident. Straight lines likewise assumed an exceptional part in Euclid's Optics as the ways followed by light beams. To logicians, for example, the German Immanuel Kant, Euclid's straight-line saying addressed a profound degree of truth. In any case, general relativity makes it conceivable deductively to look at space like some other actual amount — that is, to examine Euclid's premises. It is currently known that space-time is bended close to stars; no straight lines exist there, and light follows bended geodesics. Like Newton's law of gravity, Euclid's math accurately portrays reality under specific circumstances, however its aphorisms are not actually essential and widespread, for the universe incorporates non-Euclidean calculations also.

Taking into account its logical broadness, its reevaluating of individuals' perspective on the truth, its capacity to portray the whole universe, and its impact outside science, Einstein's relativity remains among the most critical and persuasive of logical speculations.

force, result of the mass of a molecule and its speed. Force is a vector amount; i.e., it has both extent and course. Isaac Newton's second law of movement expresses that the time pace of progress of energy is equivalent to the power following up on the molecule. See Newton's laws of movement.

From Newton's second regulation it follows that, assuming that a steady power follows up on a molecule for a given time frame, the result of power and the time stretch (the drive) is equivalent to the adjustment of the energy. On the other hand, the energy of a molecule is a proportion of the time expected for a steady power to carry it to rest.

The force of any assortment of particles is equivalent to the vector amount of the individual momenta. As per Newton's third regulation, the particles apply equivalent and inverse powers on each other, so any adjustment of the force of one molecule is precisely adjusted by an equivalent and inverse difference in the energy of another molecule. Subsequently, without any a net outside force following up on an assortment of particles, their all out energy never shows signs of change; this is the significance of the law of protection of energy. See likewise preservation regulation; precise force.

General relativity physical science

General relativity, part of the far reaching actual hypothesis of relativity framed by the German-conceived physicist Albert Einstein. It was brought about by Einstein in 1916. General relativity is worried about gravity, one of the central powers known to mankind. Gravity characterizes plainly visible way of behaving, thus broad relativity portrays huge scope actual peculiarities.

General relativity follows from Einstein's rule of identicalness: on a neighborhood scale it is difficult to recognize actual impacts because of gravity and those because of speed increase. Gravity is treated as a mathematical peculiarity that emerges from the shape of room time. The arrangement of the field conditions that depict general relativity can yield replies to various actual circumstances, like planetary elements, the birth and demise of stars, dark openings, and the development of the universe. General relativity has been tentatively confirmed by perceptions of gravitational focal points, the circle of the planet Mercury, the enlargement of time in Earth's gravitational field, and gravitational waves from blending dark openings. (For a more nitty gritty treatment of general relativity, see relativity: General relativity.)

Differential math

Differential calculation, part of math that concentrates on the calculation of bends, surfaces, and manifolds (the higher-layered analogs of surfaces). The discipline owes its name to its utilization of thoughts and procedures from differential math, however the cutting edge subject frequently utilizes logarithmic and simply mathematical strategies all things being equal. Albeit fundamental definitions, documentations, and logical depictions change broadly, the accompanying mathematical inquiries win: How can one quantify the shape of a bend inside a surface (natural) versus inside the including space (outward)? How might the curve of a surface be estimated? What is the briefest way inside a surface between two focuses on a superficial level? How is the briefest way on a surface connected with the idea of a straight line? While bends had been examined since artifact, the disclosure of analytics in the seventeenth century opened up the investigation of additional muddled plane bends — like those delivered by the French mathematician René Descartes (1596-1650) with his "compass" (see History of calculation: Cartesian math). Specifically, essential math prompted general arrangements of the old issues of tracking down the bend length of plane bends and the area of plane figures. This thusly opened the stage to the examination of bends and surfaces in space — an examination that was the beginning of differential math.

A portion of the key thoughts of differential calculation can be outlined by the strake, a spiraling strip frequently planned by designers to give underlying scaffolding to huge metal chambers like smokestacks. A strake can be framed by cutting an annular strip (the district between two concentric circles) from a level sheet of steel and afterward twisting it into a helix that twisting around the chamber, as shown in the figure. What should the sweep r of the annulus be to create the best fit? Differential calculation supplies the answer for this issue by characterizing an exact estimation for the curve of a bend; then r can be changed until the shape of within edge of the annulus matches the ebb and flow of the helix.

A significant inquiry remains: Could the annular strip at any point be bowed, without extending, so it frames a strake around the chamber? Specifically, this implies that distances estimated along the surface (natural) are unaltered. Two surfaces are supposed to be isometric in the event that one can be twisted (or changed) into the other without changing characteristic distances. (For instance, on the grounds that a piece of paper can be moved into a cylinder without extending, the sheet and cylinder are "locally" isometric — just locally in light of the fact that new, and perhaps more limited, courses are made by interfacing the two edges of the paper.) Subsequently, the subsequent inquiry becomes: Are the annular strip and the strake isometric? To respond to this and comparative inquiries, differential calculation fostered the thought of the bend of a surface.

Shape of bends

In spite of the fact that mathematicians from days of yore had portrayed a few bends as bending more than others and straight lines as not bending by any means, it was the German mathematician Gottfried Leibniz who, in 1686, first characterized the shape of a bend at each point as far as the circle that best approximates the bend by then. Leibniz named his approximating circle (as displayed in the figure) the kissing circle, from the Latin osculate ("to kiss"). He then characterized the arch of the bend (and the circle) as 1/r, where r is the sweep of the kissing circle. As a bend becomes straighter, a circle with a bigger span should be utilized to surmise it, thus the subsequent shape diminishes. In the breaking point, a straight line is supposed to be identical to a circle of limitless range and its shape characterized as zero all over the place. The main bends in normal Euclidean space with consistent curve are straight lines, circles, and helices. Practically speaking, ebb and flow is found with an equation that gives the pace of progress, or subordinate, of the digression to the bend as one maneuvers along the bend. This recipe was found by Isaac Newton and Leibniz for plane bends in the seventeenth hundred years and by the Swiss mathematician Leonhard Euler for bends in space in the eighteenth 100 years. (Note that the subsidiary of the digression to the bend isn't equivalent to the subsequent subordinate concentrated on in math, which is the pace of progress of the digression to the bend as one action along the x-hub.)

With these definitions set up, it is currently conceivable to process the ideal inward sweep r of the annular strip that goes into making the strake displayed in the figure. The annular strip's internal curve 1/r should approach the ebb and flow of the helix on the chamber. Assuming R is the span of the chamber and H is the level of one turn of the helix, then the curve of the helix is $4\pi2R/[H2 + (2\pi R)2]$. For instance, in the event that R = 1 meter and H = 10 meters, r = 3.533 meters.

Shape of surfaces

To gauge the bend of a surface at a point, Euler, in 1760, saw cross segments of the surface made via planes that contain the line opposite (or "typical") to the surface at the point (see figure). Euler called the curves of these cross segments the typical arches of the surface at the point. For instance, on a right chamber of span r, the upward cross segments are straight lines and subsequently have zero bend; the flat cross segments are circles, which have ebb and flow 1/r. The ordinary shapes at a point on a surface are by and large divergent every which way. The most extreme and least typical ebbs and flows at a point on a surface are known as the head (ordinary) curves, and the bearings wherein these typical shapes happen are known as the chief headings. Euler demonstrated that for most surfaces where the typical ebbs and flows are not steady (for instance, the chamber), these chief bearings are opposite to one another. (Note that on a circle every one of the ordinary curves are something similar and hence all are head curves.) These essential typical shapes are a proportion of how "surprising" the surface is.

The hypothesis of surfaces and head ordinary curves was broadly evolved by French geometers drove by Gaspard Monge (1746-1818). It was in an 1827 paper; in any case, that the German mathematician Carl Friedrich Gauss made the large advancement that permitted differential calculation to address the inquiry raised above of whether the annular strip is isometric to the strake. The Gaussian ebb and flow of a surface at a point is characterized as the result of the two head typical curves; it is supposed to be positive in the event that the central ordinary shapes bend in a similar course and negative assuming they bend in inverse headings. Ordinary bends for a plane surface are each of the zero, and in this way the Gaussian ebb and flow of a plane is zero. For a chamber of span r, the base typical shape is zero (along the upward straight lines), and the most extreme is 1/r (along the even circles). In this manner, the Gaussian ebb and flow of a chamber is likewise zero.

On the off chance that the chamber is cut along one of the upward straight lines, the subsequent surface can smoothed (without stretch) onto a square shape. In differential math, it is said that the plane and chamber are locally isometric. These are extraordinary instances of two significant hypotheses:

The Gaussian arch of an annular strip (being in the plane) is continually zero. So to answer whether the annular strip is isometric to the strake, one necessities just to check whether a strake has steady zero Gaussian curve. The Gaussian curve of a strake is really bad, thus the annular strip should be extended — albeit this can be limited by restricting the shapes.

Most limited ways on a surface

From an outside, or outward, viewpoint, no bend on a circle is straight. By and by, the incredible circles are naturally straight — an insect slithering along an extraordinary circle doesn't turn or bend regarding the surface. Around 1830 the Estonian mathematician Ferdinand Disapproving of characterized a bend on a surface to be a geodesic assuming that it is naturally straight — that is, assuming that there is no recognizable shape from inside the surface. A significant undertaking of differential math is to decide the geodesics on a surface. The incredible circles are the geodesics on a circle.

An extraordinary circle curve that is longer than a half circle is naturally straight on the circle; however it isn't the briefest distance between its endpoints. Then again, the briefest way in a surface isn't generally straight, as displayed in the figure. A significant hypothesis is:

On a surface which is finished (each geodesic can be expanded endlessly) and smooth, each briefest bend is naturally straight and each characteristically straight bend is the most limited bend between neighboring focuses.

Differential calculation rundown

Differential calculation, Field of science in which techniques for math are applied to the nearby calculation of bends and surfaces (i.e., to a little part of a surface or bend around a point). A basic model is finding the digression line on a two-layered bend at a given point. Comparative tasks might be reached out to compute the shape and length of a bend and to closely resembling properties of surfaces in quite a few aspects.

Differential calculation is a numerical discipline that concentrates on the math of smooth shapes and smooth spaces, also called smooth manifolds. It utilizes the procedures of differential math, necessary analytics, direct polynomial math and multilinear polynomial math. The field has its beginnings in the investigation of round calculation as far back as vestige. It likewise connects with stargazing, the geodesy of the Earth, and later the investigation of exaggerated calculation by Lobachevski. The least complex instances of smooth spaces are the plane and space bends and surfaces in the three-layered Euclidean space, and the investigation of these shapes framed the reason for advancement of present day differential calculation during the eighteenth and nineteenth hundreds of years.

Since the late nineteenth 100 years, differential math has developed into a field concerned all the for the most part with mathematical designs on differentiable manifolds. A mathematical construction is one which characterizes some idea of size, distance, shape, volume, or other rigidifying structure. For instance, in Riemannian math distances and points are determined, in simplistic calculation volumes might be registered, in conformal math just points are determined, and in measure hypothesis certain fields are given over the space. Differential math is firmly connected with, and is in some cases taken to incorporate, differential geography, which frets about properties of differentiable manifolds which depend on no extra mathematical construction (see that article for more conversation on the qualification between the two subjects). Differential calculation is likewise connected with the mathematical parts of the hypothesis of differential conditions, also called mathematical examination.

Differential calculation tracks down applications all through arithmetic and the innate sciences. Most noticeably the language of differential math was utilized by Albert Einstein in his hypothesis of general relativity, and accordingly by physicists in the advancement of quantum field hypothesis and the standard model of molecule physical science. Beyond material science, differential calculation tracks down applications in science, financial aspects, designing, control hypothesis, PC illustrations and PC vision, and as of late in AI.

Old style artifact until the Renaissance (300 BC - 1600 Promotion)

The investigation of differential calculation, or if nothing else the investigation of the math of smooth shapes, can be followed back basically to old style vestige. Specifically, a lot was had some significant awareness of the calculation of the Earth, a round math, in the hour of the old Greek mathematicians. Broadly, Eratosthenes determined the perimeter of the Earth around 200 BC, and around 150 Promotion Ptolemy in his Geology presented the stereographic projection for the motivations behind planning the state of the Earth.[1] Verifiably over the course of this time rules that structure the groundwork of differential math and analytics were utilized in geodesy, albeit in a much worked on structure. In particular, as far back as Euclid's Components it was perceived that a straight line could be characterized by its property of giving the briefest distance between two focuses, and applying this equivalent standard to the outer layer of the Earth prompts the end that extraordinary circles, which are simply locally like straight lines in a level plane, give the most limited way between two focuses on the World's surface. For sure the estimations of distance along such geodesic ways by Eratosthenes and others can be viewed as a simple proportion of arc length of bends, an idea which didn't see a thorough definition as far as math until the 1600s.

Close to this time there were just insignificant clear utilizations of the hypothesis of infinitesimals to the investigation of math, a forerunner to the cutting edge math based investigation of the subject. In Euclid's Components the idea of juncture of a line to a circle is examined, and Archimedes applied the technique for fatigue to register the areas of smooth shapes like the circle, and the volumes of smooth three-layered solids like the circle, cones, and cylinders.

There was little advancement in the hypothesis of differential calculation among relic and the start of the Renaissance. Before the improvement of math by Newton and Leibniz, the main advancement in the comprehension of differential calculation came from saving guide of the World's surface onto a level plane, a result of the later Theorem Egregious of Gauss.

The primary efficient or thorough treatment of calculation utilizing the hypothesis of infinitesimals and ideas from math started around the 1600s when analytics was first evolved by Gottfried Leibniz and Isaac Newton. Right now, the new work of René Descartes acquainting insightful directions with math permitted mathematical states of expanding intricacy to be portrayed thoroughly. Specifically close to this time Pierre de Fermat, Newton, and Leibniz started the investigation of plane bends and the examination of ideas, for example, places of enunciation and circles of endearment, which help in the estimation of ebb and flow. Without a doubt currently in his most memorable paper on the underpinnings of math, Leibniz noticed that the minuscule condition.

Around this equivalent time, Leonhard Euler, initially an understudy of Johann Bernoulli, gave numerous huge commitments to the improvement of calculation, however to science more broadly.[3] concerning differential math, Euler concentrated on the thought of a geodesic on a surface determining the principal scientific geodesic condition, and later presented the primary arrangement of natural direction frameworks on a surface, starting the hypothesis of characteristic calculation whereupon current mathematical thoughts are based.[1] Close to this time Euler's investigation of mechanics in the Mechanical lead to the acknowledgment that a mass going along a surface not under the impact of any power would navigate a geodesic way, an early forerunner to the significant fundamental thoughts of Einstein's overall relativity, and furthermore to the Euler-Lagrange conditions and the main hypothesis of the analytics of varieties, which supports in present day differential calculation numerous strategies in simplistic math and mathematical examination. This hypothesis was utilized by Lagrange, a co-engineer of the math of varieties, to determine the primary differential condition depicting a negligible surface regarding the Euler-Lagrange condition. In 1760 Euler demonstrated a hypothesis communicating the bend of a space bend based on a surface in conditions of the essential curves, known as Euler's hypothesis.

Later during the 1700s, the new French school drove by Gaspard Monge started to make commitments to differential calculation. Monge made significant commitments to the hypothesis of plane bends, surfaces, and concentrated on surfaces of upset and envelopes of plane bends and space bends. A few understudies of Monge made commitments to this equivalent hypothesis, and for instance Charles Dupin gave another understanding of Euler's hypothesis.

Polarization physical science

Polarization, property of specific electromagnetic radiations in which the bearing and greatness of the vibrating electric field are connected in a predefined way.

Light waves are cross over: that is, the vibrating electric vector related with each wave is opposite to the bearing of proliferation. A light emission light comprises of waves moving in similar course with their electric vectors pointed in irregular directions about the pivot of engendering. Plane captivated light comprises of waves in which the course of vibration is no different for all waves. In round polarization the electric vector turns about the bearing of proliferation as the wave advances. Light might be enraptured by reflection or by going it through channels, for example, certain precious stones, that communicate vibration in one plane yet not in others.

Atomic response, change in the personality or attributes of a nuclear core, prompted by assaulting it with a vigorous molecule. The barraging molecule might be an alpha molecule, a gamma-beam photon, a neutron, a proton, or a weighty particle. Regardless, the barraging molecule should have sufficient energy to move toward the decidedly charged core to close enough to serious areas of strength for the power.

A normal atomic response includes two responding particles — a weighty objective core and a light barraging molecule — and produces two new particles — a heavier item core and a lighter catapulted molecule. In the main noticed atomic response (1919), Ernest Rutherford barraged nitrogen with alpha particles and recognized the launched out lighter particles as hydrogen cores or protons ($^{1}_{1}H$ or p) and the item cores as an uncommon oxygen isotope. In the principal atomic response delivered by falsely sped up particles (1932), the English physicists J.D. Cockcroft and E.T.S. Walton barraged lithium with sped up protons and consequently created two helium cores, or alpha particles. As it has become conceivable to speed up charged particles to progressively more noteworthy energy, some high-energy atomic responses have been seen that produce different subatomic particles called mesons, baryons, and reverberation particles. Logical calculation, likewise called coordinate math, numerical subject in which arithmetical imagery and strategies are utilized to address and take care of issues in math. The significance of logical math is that it lays out a correspondence between mathematical bends and logarithmic conditions. This correspondence makes it conceivable to reformulate issues in calculation as comparable issues in variable based math, as well as the other way around; the strategies for either subject can then be utilized to tackle issues in the other. For instance, PCs make activities for show in games and movies by controlling logarithmic conditions.

Rudimentary insightful math

Apollonius of Pergo (c. 262-190 BC), referred to by his peers as the "Incomparable Geometer," foreshadowed the advancement of insightful calculation by over 1,800 years with his book Conics. He characterized a conic as the convergence of a cone and a plane (see figure). Utilizing Euclid's outcomes on comparative triangles and on secants of circles, he found a connection fulfilled by the good ways from any guide P of a conic toward two opposite lines, the significant hub of the conic and the digression at an endpoint of the hub. These distances compare to directions of P, and the connection between these directions relates to a quadratic condition of the conic. Apollonius utilized this connection to conclude key properties of conics. See conic area.

Further advancement of direction frameworks (see add) in arithmetic arose solely after polynomial math had developed under Islamic and Indian mathematicians. (See arithmetic: The Islamic world (eighth fifteenth hundreds of years) and math, South Asian.) Toward the finish of the sixteenth 100 years, the French mathematician François Viète presented the principal efficient logarithmic documentation, utilizing letters to address known and obscure mathematical amounts, and he grew strong general strategies for working with mathematical articulations and tackling mathematical conditions. With the force of arithmetical documentation, mathematicians were at this point not totally reliant upon mathematical figures and mathematical instinct to take care of issues. The more trying started to abandon the standard mathematical perspective in which straight (first power) factors compared to lengths, squares (second capacity) to regions, and cubic (third capacity) to volumes, with higher powers lacking "physical" translation. Two Frenchmen, the mathematician-savant René Descartes and the legal advisor mathematician Pierre de Fermat, were among quick to make this thinking for even a moment to stride.

Descartes and Fermat autonomously established logical calculation during the 1630s by adjusting Viète's polynomial math to the investigation of mathematical loci. They moved unequivocally past Viète by utilizing letters to address removes that are variable rather than fixed. Descartes utilized conditions to concentrate on bends characterized mathematically, and he focused on the need to think about broad logarithmic bends — diagrams of polynomial conditions in x and y, all things considered. He showed his strategy on a traditional issue: finding all focuses P to such an extent that the result of the good ways from P to specific lines rises to the result of the distances to different lines. See calculation: Cartesian math. Fermat underlined that any connection among x and y organizes decides a bend (see figure). Utilizing this thought, he recast

Apollonius' contentions in arithmetical terms and reestablished lost work. Fermat showed that any quadratic condition in x and y can be placed into the standard type of one of the conic segments.

Fermat didn't distribute his work, and Descartes purposely made his hard to peruse to deter "dilettantes." Their thoughts acquired general acknowledgment just through the endeavors of different mathematicians in the last 50% of the seventeenth hundred years. Specifically, the Dutch mathematician Franz van Scooted deciphered Descartes' compositions from French to Latin. He added indispensable informative material, as did the French attorney Florimond de Beaune, and the Dutch mathematician Johan de Witt. In Britain, the mathematician John Wallis promoted scientific calculation, utilizing conditions to characterize conics and determine their properties. He utilized negative facilitates unreservedly, in spite of the fact that it was Isaac Newton who unequivocally utilized two (angled) tomahawks to separate the plane into four quadrants, as displayed in the figure. Insightful calculation greatest affected math by means of analytics. Without admittance to the force of insightful math, old style

Greek mathematicians like Archimedes (c. 285-212/211 BC) tackled unique instances of the fundamental issues of math: tracking down digressions and outrageous focuses (differential analytics) and bend lengths, regions, and volumes (necessary math). Renaissance mathematicians were driven back to these issues by the requirements of stargazing, optics, route, fighting, and trade. They normally tried to utilize the force of polynomial math to characterize and examine a developing scope of bends. Fermat fostered a mathematical calculation for finding the digression to a logarithmic bend at a point by finding a line that has a twofold convergence with the bend at the point — generally, designing differential analytics. Descartes presented a comparable yet more confounded calculation utilizing a circle. Fermat registered regions under the bends y = axk for all judicious numbers k ≠ −1 by adding areas of engraved and encompassed square shapes. (See weariness, strategy for.) Until the end of the seventeenth 100 years, the preparation for analytics was gone on by numerous mathematicians, including the Frenchman Gilles Personnel de Roberta, the

Italian Bonaventura Cavalier, and the Britons James Gregory, John Wallis, and Isaac Pushcart.

Newton and the German Gottfried Leibniz altered arithmetic toward the finish of the seventeenth hundred years by freely exhibiting the force of analytics. The two men utilized directions to foster documentations that communicated the thoughts of analytics in full consensus and driven normally to separation rules and the major hypothesis of math (associating differential and fundamental math). See examination.

Newton showed the significance of scientific strategies in calculation, aside from their part in analytics, when he stated that any cubic — or, logarithmic bend of degree three — has one of four standard conditions,

$$xy2 + ey = ax3 + bx2 + cx + d,$$

$$xy = ax3 + bx2 + cx + d,$$

$$y2 = ax3 + bx2 + cx + d,$$

$$y = ax3 + bx2 + cx + d,$$

For appropriate direction tomahawks. The Scottish mathematician James Sterling demonstrated this statement in 1717, perhaps with Newton's guide. Newton separated cubic into 72 species, an all-out later rectified to 78.

Newton likewise told the best way to communicate a mathematical bend close to the beginning regarding the partial power series $y = a_1 x^{1/k} + a_2 x^{2/k} + \ldots$ for a positive whole number k. Mathematicians have since utilized this method to concentrate on arithmetical bends, all things considered.

Insightful calculation of three and more aspects

Albeit both Descartes and Fermat recommended utilizing three directions to concentrate on bends and surfaces in space, three-layered scientific calculation grew gradually until around 1730, when the Swiss mathematicians Leonhard Euler and Jacob Hermann and the French mathematician Alexis Claimant created general conditions for chambers, cones, and surfaces of upset. For instance, Euler and Hermann showed that the condition $f(z) = x^2 + y^2$ gives the surface that is created by rotating the bend $f(z) = x2$ about the z-hub (see the figure, which shows the elliptic parabolic $z = x2 + y2$).

Newton made the amazing case that all plane cubics emerge from those in his third standard structure by projection between planes. This was demonstrated autonomously in 1731 by Clairaut and the French mathematician François Nicole. Clairaut got all the cubics in Newton's four standard structures as areas of the cubical cone

$$zy2 = ax3 + bx2z + cxz2 + dz3$$

In 1843 the Irish mathematician-cosmologist William Rowan Hamilton addressed four-layered vectors logarithmically and imagined the quaternions, the primary noncommutative polynomial math to be broadly contemplated. Duplicating quaternions with one direction zero drove Hamilton to find principal procedure on vectors. By the by, numerical physicists found the documentation utilized in vector examination more adaptable — specifically, it is promptly extendable to boundless layered spaces. The quaternions survived from interest logarithmically and were integrated during the 1960s into specific new molecule physical science models.

As promptly accessible figuring power filled dramatically somewhat recently of the twentieth 100 years, PC movement and PC supported plan became universal. These applications depend on three-layered insightful math. Arranges are utilized to decide the edges or parametric bends that structure limits of the surfaces of virtual items. Vector investigation is utilized to demonstrate lighting and decide practical shades of surfaces.

As soon as 1850, Julius Pluckier had joined logical and projective math by presenting homogeneous directions that address focuses in the Euclidean plane (see Euclidean calculation) and at endlessness in a uniform way as triples. Projective changes, which are invertible straight changes of homogeneous directions, are given by lattice duplication. This lets PC illustrations programs productively change the shape or the perspective on imagined items and undertaking them from three-layered virtual space to the two-layered survey screen.

Non-Euclidean calculation

Non-Euclidean calculation, in a real senses any math that isn't equivalent to Euclidean calculation. Albeit the term is regularly used to allude just to exaggerated math, normal use incorporates those couple of calculations (exaggerated and circular) that vary from however are exceptionally near Euclidean calculation (see table).

The non-Euclidean calculations created along two different verifiable strings. The main string began with the hunt to figure out the development of stars and planets in the clearly hemispherical sky. For instance, Euclid (prospered c. 300 BCE) expounded on circular calculation in his galactic work Phaenomena. As well as focusing on the sky, the people of yore endeavored to grasp the state of the Earth and to involve this comprehension to take care of issues in route over significant distances (and later for enormous scope reviewing). These exercises are parts of circular calculation.

For a considerable length of time following Euclid, mathematicians endeavored either to demonstrate the hypothesize as a hypothesis (in view of different proposes) or to change it in different ways. (All see calculation: Non-Euclidean calculations.) These endeavors finished when the Russian Nikolay Lobachevsky (1829) and the Hungarian János Bolyai (1831) freely distributed a depiction of a math that, with the exception of the equal propose, fulfilled Euclid's hypothesizes and normal thoughts. This calculation is called exaggerated math.

Circular math

From early times, individuals saw that the briefest distance between two focuses on Earth were incredible circle courses. For instance, the Greek stargazer Ptolemy wrote in Geology (c. 150 CE):

Incredible circles are the "straight lines" of circular math. This is a result of the properties of a circle, where the most limited distances on a superficial level are incredible circle courses. Such bends are supposed to be "naturally" straight. (Note, in any case, that naturally straight and most limited are not really indistinguishable, as displayed in the figure.) Three meeting extraordinary circle curves structure a circular triangle (see figure); while a round triangle should be contorted to fit on one more circle with an alternate span, the thing that matters is only one of scale. In differential math, circular calculation is portrayed as the math of a surface with steady certain curve.

There are numerous approaches to extending a part of a circle, like the outer layer of the Earth, onto a plane. These are known as guides or graphs and they should essentially twist distances and either region or points. Map makers' requirement for different characteristics in map projections gave an early stimulus to the investigation of circular calculation.

Elliptic math is the term used to show a proverbial formalization of circular calculation in which each sets of antipodal focuses is treated as a solitary point. A characteristic scientific perspective on round calculation was created in the nineteenth 100 years by the German mathematician Bernhard Riemann; as a rule called the Riemann circle (see figure), it is concentrated on in college seminars on complex examination. A few texts call this (and hence round calculation) Riemannian math, yet this term all the more accurately applies to a piece of differential math that gives an approach to naturally portraying any surface.

Exaggerated calculation

The main portrayal of exaggerated math was given with regards to Euclid's hypothesizes, and it was before long demonstrated that all exaggerated calculations vary just in scale (in the very sense that circles just contrast in size). During the nineteenth century it was shown that exaggerated surfaces should have steady regrettable ebb and flow. In any case, this actually left open whether or not any surface with exaggerated calculation really exists.

In 1868 the Italian mathematician Eugenio Beltrami depicted a surface, called the pseudosphere that has consistent negative ebb and flow. In any case, the pseudosphere is certainly not a total model for exaggerated math, in light of the fact that characteristically straight lines on the pseudosphere may cross themselves and can't be gone on past the jumping circle (neither of which is valid in exaggerated calculation). In 1901 the German mathematician David Hilbert demonstrated that it is difficult to characterize a total exaggerated surface utilizing truly logical capabilities (basically, capabilities that can be communicated with regards to common recipes). Back then, a surface generally implied one characterized by truly scientific capabilities, thus the inquiry was deserted. Nonetheless, in 1955 the Dutch mathematician Nicolaas Kuiper demonstrated the presence of a total exaggerated surface, and during the 1970s the American mathematician William Thurston portrayed the development of an exaggerated surface. Such a surface, as displayed in the figure, can likewise be sewn.

In the nineteenth hundred years, mathematicians created three models of exaggerated calculation that can now be deciphered as projections (or guides) of the exaggerated surface. Albeit these models all experience the ill effects of some contortion — like the way that level guides mutilate the round Earth — they are helpful exclusively and in mix as assistants to grasp exaggerated calculation. In 1869-71 Beltrami and the German mathematician Felix Klein fostered the principal complete model of exaggerated calculation (and first referred to the math as "exaggerated"). In the Klein-Beltrami model (displayed in the figure, upper left), the exaggerated surface is planned to the inside of a circle, with geodesics in the exaggerated surface relating to harmonies in the circle. In this manner, the Klein-Beltrami model jam "straightness" yet at the expense of contorting points. Around 1880 the French mathematician Henri Poincaré created two additional models. In the Poincaré plate model (see figure, upper right), the exaggerated surface is planned to the inside of a round circle, with exaggerated geodesics planning to round circular segments (or breadths) in the circle that meet the jumping circle at right points. In the Poincaré upper half-plane model (see figure, base), the exaggerated surface is planned onto the half-plane over the x-pivot, with exaggerated geodesics planned to crescents (or vertical beams) that meet the x-hub at right points. Both Poincaré models misshape distances while protecting points as estimated by digression lines.

Radioactivity

Radioactivity, property displayed by specific kinds of matter of transmitting energy and subatomic particles immediately. It is, generally, a quality of individual nuclear cores.

A shaky core will disintegrate precipitously, or rot, into a steadier setup however will do so just in a couple of explicit ways by discharging specific particles or certain types of electromagnetic energy. Radioactive rot is a property of a few normally happening components as well as of misleadingly delivered isotopes of the components. The rate at which a radioactive component rots is communicated with regards to its half-life; i.e., the time expected for one-half of some random amount of the isotope to rot. Half-life range from over 1024 years for certain cores to under 10−23 second (see beneath Paces of radioactive changes). The result of a radioactive rot process — called the girl of the parent isotope — may itself be temperamental, in which case it, as well, will rot. The interaction goes on until a stable nuclide has been shaped.

The idea of radioactive outflows

The outflows of the most well-known types of unconstrained radioactive rot are the alpha (α) molecule, the beta (β) molecule, the gamma (γ) beam, and the neutrino. The alpha molecule is really the core of a helium-4 iota, with two positive charges

He. Such charged particles are called particles. The nonpartisan helium particle has two electrons outside its core adjusting these two charges. Beta particles might be adversely charged (beta short, image e−), or decidedly charged (beta furthermore, image e+). The beta short [$\beta-$] molecule is really an electron made in the core during beta rot with practically no relationship to the orbital electron haze of the iota. The beta in addition to molecule, likewise called the positron, is the antiparticle of the electron; when united; two such particles will commonly destroy one another. Gamma beams are electromagnetic radiations like radio waves, light, and X-beams. Beta radioactivity additionally creates the neutrino and antineutrino, particles that have no charge and very little mass, symbol Kinds of radioactivity

The early work on normal radioactivity related with uranium and thorium metals recognized two unmistakable kinds of radioactivity: alpha and beta rot. In alpha rot, a fiery helium particle (alpha molecule) is launched out, leaving a girl core of nuclear number two not exactly the parent and of nuclear mass number four not exactly the parent. A model is the rot (represented by a bolt) of the bountiful isotope of uranium, 238U, to a thorium girl in addition to an alpha molecule:

Given for this and resulting responses are the energy delivered (Q) in huge number of electron volts (MeV) and the half-life (t1⁄2). It ought to be noticed that in alpha rots the charges, or number of protons, displayed in addendum are in balance on the two sides of the bolt, similar to the nuclear masses, displayed in superscript.

Portrayal of the results of a radioactive rot.

In the above response for beta rot, v addresses the antineutrino. Here, the quantity of protons is expanded by one in the response, however the complete charge continues as before, on the grounds that an electron, with negative charge, is likewise made.

Gamma rot

A third sort of radiation, gamma radiation, generally goes with alpha or beta rots. Gamma beams are photons and are without rest mass or charge. Alpha or beta rot may just continue straightforwardly to the ground (least energy) condition of the little girl core without gamma discharge, however the rot may likewise continue completely or part of the way to higher energy states (invigorated conditions) of the girl. In the last option case, gamma discharge might happen as the energized states change to bring down energy conditions of a similar core. (On the other hand to gamma outflow, an invigorated core might change to a lower energy state by catapulting an electron from the cloud encompassing the core. This orbital electron launch is known as inward change and leads to an enthusiastic electron and frequently a X-beam as the nuclear cloud fills in the void orbital of the shot out electron. The proportion of inward transformation to the elective gamma emanation is known as the inside change coefficient.)

Isomeric changes

There is a great many paces of half-lives for the gamma-emanation process. Normally dipole advances (see underneath Gamma progress), in which the gamma beam takes away one $\hbar$ unit of precise energy, are quick, not as much as nanoseconds (one nanosecond approaches $10-9$ second). The law of protection of rakish energy expects that the amount of precise momenta of the radiation and little girl core is equivalent to the rakish force (turn) of the parent. On the off chance that the twists of starting and last states vary by more than one, dipole radiation is prohibited, and gamma discharge should continue all the more leisurely by a, the energized core is supposed to be in a metastable, or isomeric, express (the names for an enduring invigorated state), and characterizing the rot as one more sort of radioactivity, an isomeric transition is standard. An illustration of isomerism is found in the protactinium-234 core of the uranium-238 rot chain:

The letter m following the mass number represents metastable and shows an atomic isomer.

During the 1930s new kinds of radioactivity were found among the fake results of atomic responses: beta-in addition to rot, or positron discharge, and electron catch. In beta-in addition to rot a fiery positron is made and produced, alongside a neutrino, and the core changes to a little girl, lower by one in nuclear number and a similar in mass number. For example, carbon-11 (Z = 6) rots to boron-11 (Z = 5), in addition to one positron and one neutrino:

Electron catch (EC) is a cycle where rot follows the catch by the core of an orbital electron. It is like positron rot in that the core changes to a girl of one lower nuclear number. It contrasts in that an orbital electron from the cloud is caught by the core with resulting discharge of a nuclear X-beam as the orbital opportunity is filled by an electron from the cloud about the core. A model is the core of beryllium-7 catching one of its inward electrons to give lithium-7:

The fundamental elements of radioactive rot of atomic animal varieties are in many cases shown in a rot plot. Figure 1 shows the rot plan of beryllium-7.

Shown are the half-existence of the parent and that of the energized girl state, as well as its energy 0.4774 MeV. The twists and equalities of every one of the three states are given on the upper left-hand side of upward bolt representing the gamma change. The skewed bolts represent the electron-catch rot with marks giving the level of rot straightforwardly to ground state (89.7 percent) and the level of EC rot going by means of the energized state (10.3 percent). The boldface numbers following the rates are supposed log ft values, to be experienced underneath regarding beta-rot rates. The general energy discharge, QEC, is shown beneath. The QEC is essentially a determined worth since there is no broad down to earth method for estimating the neutrino energies going with EC rot. With a couple of electron-catching nuclides, it has been feasible to quantify straightforwardly the rot energy by estimation of an intriguing interaction called internal bremsstrahlung (slowing down radiation). In this cycle the energy discharge is divided among the neutrino and a gamma beam. The deliberate dissemination of gamma-beam energies demonstrates the absolute energy discharge. Ordinarily there is such a lot of normal gamma radiation with radioactive rot that the internal bremsstrahlung is inconspicuous.

Unconstrained splitting

One more kind of radioactivity is unconstrained splitting. In this cycle the core parts into two piece cores of generally around 50% of the mass of the parent. This cycle is just scarcely discernible in rivalry with the more common alpha rot for uranium, yet for the absolute heaviest fake cores, for example, fermium-256, unconstrained splitting turns into the transcendent method of radioactive rot. Dynamic energy sets free from 150 to 200 MeV might happen as the sections are sped up separated by the enormous electrical repugnance between their atomic charges. The response is as per the following:

Only one of a few item sets is shown. A couple of neutrons are constantly discharged in parting of this isotope, a component crucial for chain responses. Unconstrained splitting isn't to be mistaken for actuated parting, the cycle engaged with atomic reactors. Prompted fission is a property of uranium-

235, plutonium-239, and different isotopes to go through splitting after retention of a sluggish neutron. Other than the necessity of a neutron catch to start it, initiated splitting is very like unconstrained parting in regards to add up to energy discharge, quantities of optional neutrons, etc. (see atomic parting).

Proton radioactivity

Proton radioactivity, found in 1970, is shown by an energized isomeric condition of cobalt-53, 53mCo, 1.5 percent of which discharges protons:

high energized condition of the girl core, and this state quickly transmits a weighty particle.(Note: the reference mark signifies the fleeting transitional invigorated conditions of oxygen-17, and Amax n means the most extreme energy noticed for radiated neutrons.) There is a little creation of postponed neutron producers following atomic splitting, and these radio activities are particularly significant in giving a sensible reaction time to permit control of atomic splitting reactors by precisely moved control bars. Among the positron producers in the light-component locale, a number beta rot mostly to energized states that are unsteady concerning emanation of an alpha molecule. Hence, these species show alpha radiation with the half-existence of the beta outflow. Both the positron rot from boron-8 and electron rot from lithium-8 are beta-deferred alpha outflow, since ground as well as invigorated conditions of beryllium-8 are unsteady regarding separation into two alpha particles. Another model, sodium-20 (20Na) to give progressively neon-20 (20Ne; the bullet again demonstrating the fleeting halfway state) lastly oxygen-16 is recorded beneath:

Weighty particle radioactivity

In 1980 A. Sandulescu, D.N. Poenaru, and W. Greiner depicted estimations showing the chance of another kind of rot of weighty cores halfway between alpha rot and unconstrained parting. The main perception of weighty particle radioactivity was that of a 30-MeV, carbon-14 outflow from radium-223 by H.J. Rose and G.A. Jones in 1984. The proportion of carbon-14 rot to alpha rot is around 5 × 10−10. Perceptions likewise have been made of carbon-14 from radium-222, radium-224, and radium-226, as well as neon-24 from thorium-230, protactinium-231, and uranium-232. Such weighty particle radioactivity, similar to alpha rot and unconstrained splitting, includes quantum-mechanical burrowing through the potential-energy boundary. Shell impacts assume a significant part in this peculiarity and in all cases saw to date the weighty accomplice of carbon-14 or neon-24 is near doubly wizardry lead-208 (see underneath Atomic models).

Event of radioactivity

A few types of radioactivity happen normally on the planet. A couple of animal categories have half-life equivalent to the age of the components (around 6 × 109 years), with the goal that they have not rotted away after their development in stars. Eminent among these are uranium-238, uranium-235, and thorium-232. Additionally, there is potassium-40, the central wellspring of illumination of the body through its presence in potassium of tissue. Unimportant are the beta producer's vanadium-50, rubidium-87, indium-115, tellurium-123, lanthanum-138, lutetium-176, and rhenium-187, and the alpha producers cerium-142, neodymium-144, samarium-147, gadolinium-152, dysprosium-156, hafnium-174, platinum-190, and lead-204. Other than these around 109-year species, there are the more limited lived girl exercises took care of by some of the above species; e.g., by different cores of the components between lead (Z = 82) and thorium (Z = 90).

One more classification of regular radioactivity incorporates species created in the upper environment by astronomical beam siege. Eminent are 5,720-year carbon-14 and 12.3-year tritium (hydrogen-3), 53-day beryllium-7, and 2,700,000-year beryllium-10. Shooting stars are found to contain extra limited quantities of radioactivity, the aftereffect of enormous beam bombardments during their set of experiences outside the World's environmental safeguard. Exercises as fleeting as 35-day argon-37 have been estimated in new falls of shooting stars. Atomic blasts beginning around 1945 have infused extra radioactivities into the climate, comprising of both atomic splitting items and optional items framed by the activity of neutrons from atomic weapons on encompassing matter.

The splitting items envelop the greater part of the known beta producers in the mass district 75-160. They are framed in differing yields, ascending to maxima of around 7% per splitting in the mass district 92-102 (light pinnacle of the parting yield versus nuclear mass bend) and 134-144 (weighty pinnacle). Two sorts of postponed dangers brought about by radioactivity are perceived. In the first place, the overall radiation level is raised by aftermath settling to Earth. Insurance can be given by cement or earth protecting until the movement has rotted to an adequately low level. Second, ingestion or inward breath of even low levels of specific radioactive species can represent an exceptional risk, contingent upon the half-life, nature of radiations, and synthetic conduct inside the body. For an itemized conversation on the organic impacts of radiation, see radiation: Natural impacts of ionizing radiation.